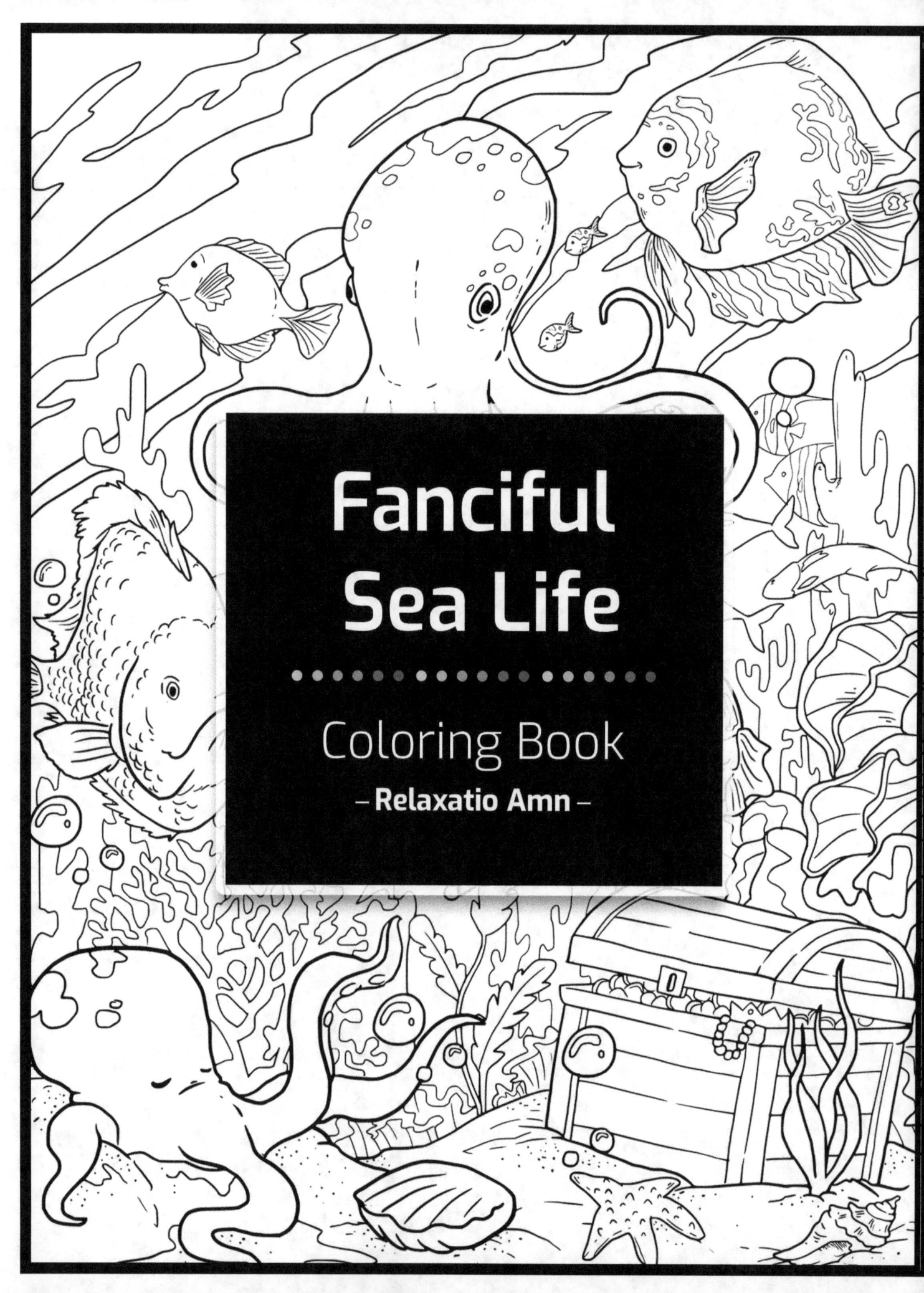

Thank you for choosing Relaxatio Amn
This coloring book make with love .

enjoy it all.

Copyright © 2020 by Relaxtion Amn

All rights reserved. This book or any portion thereof may not be reproduced or used in any manner whatsoever without the express written permission of the publisher except for the use of brief quotations in a book review.

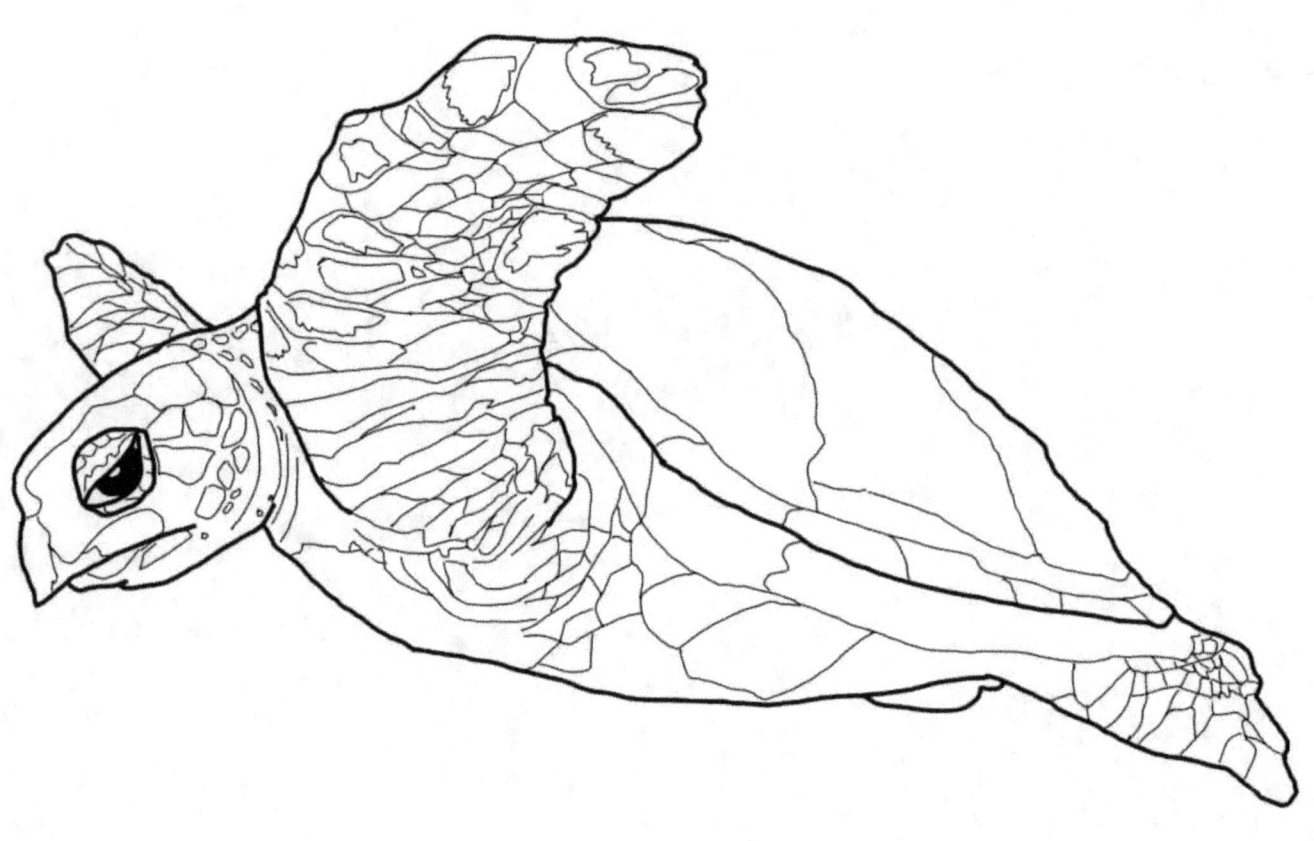

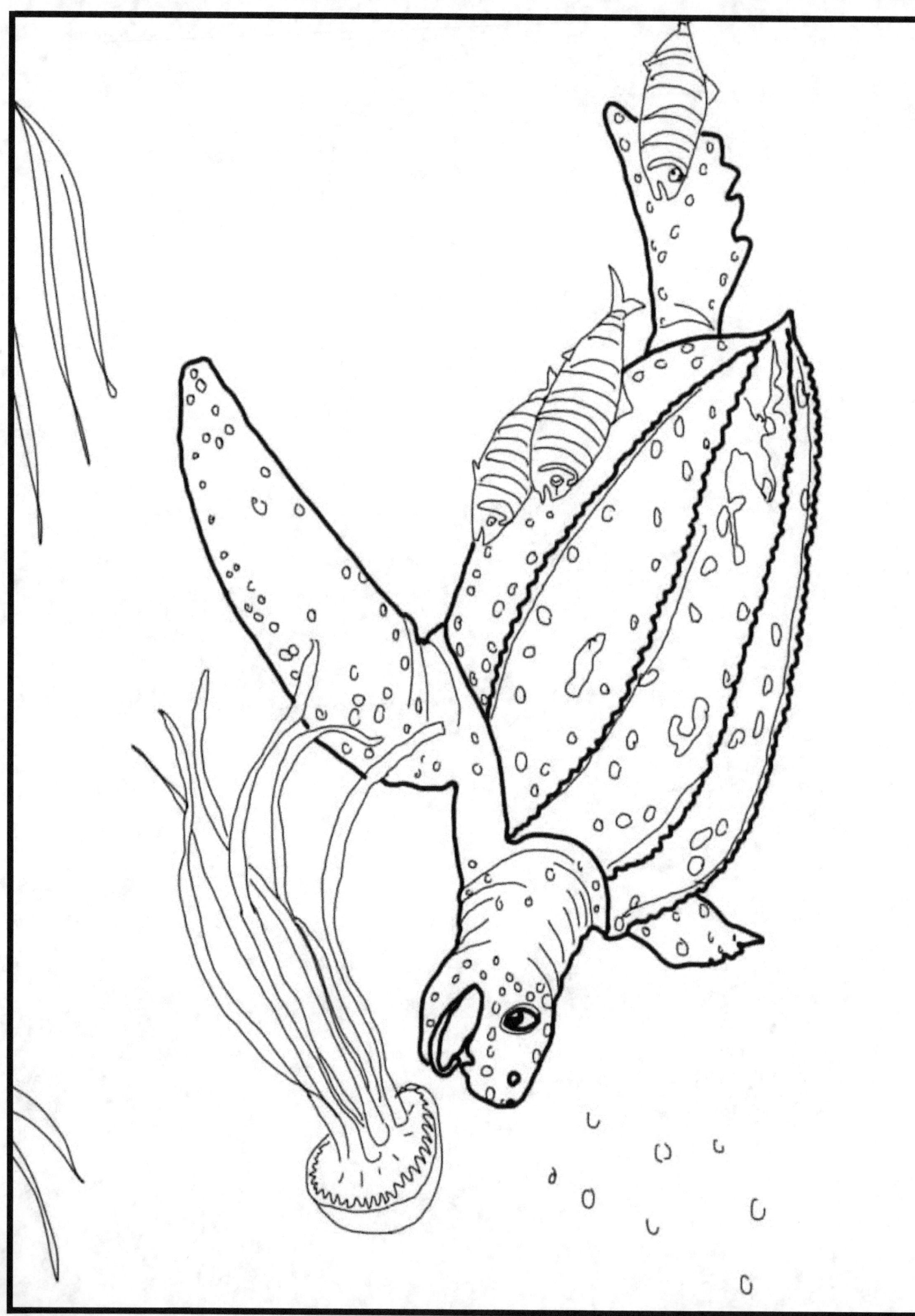

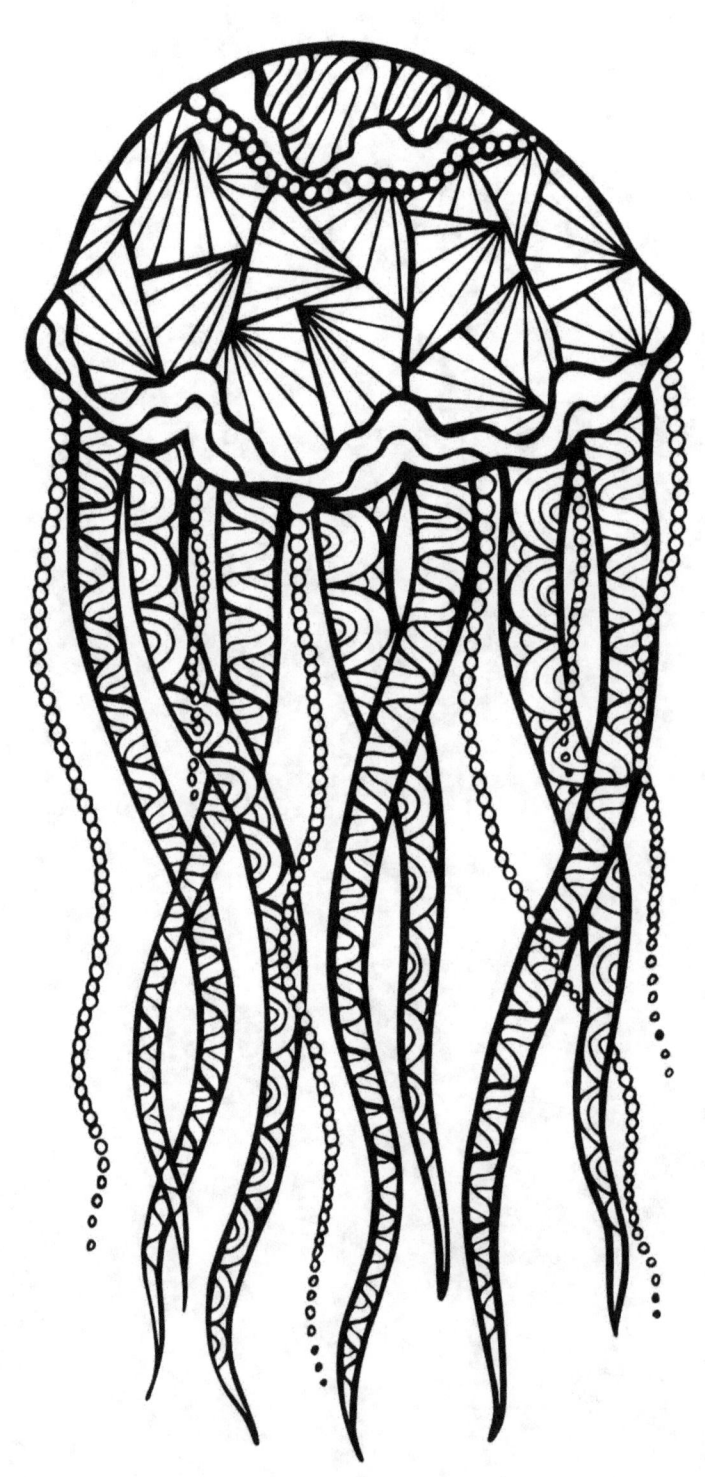

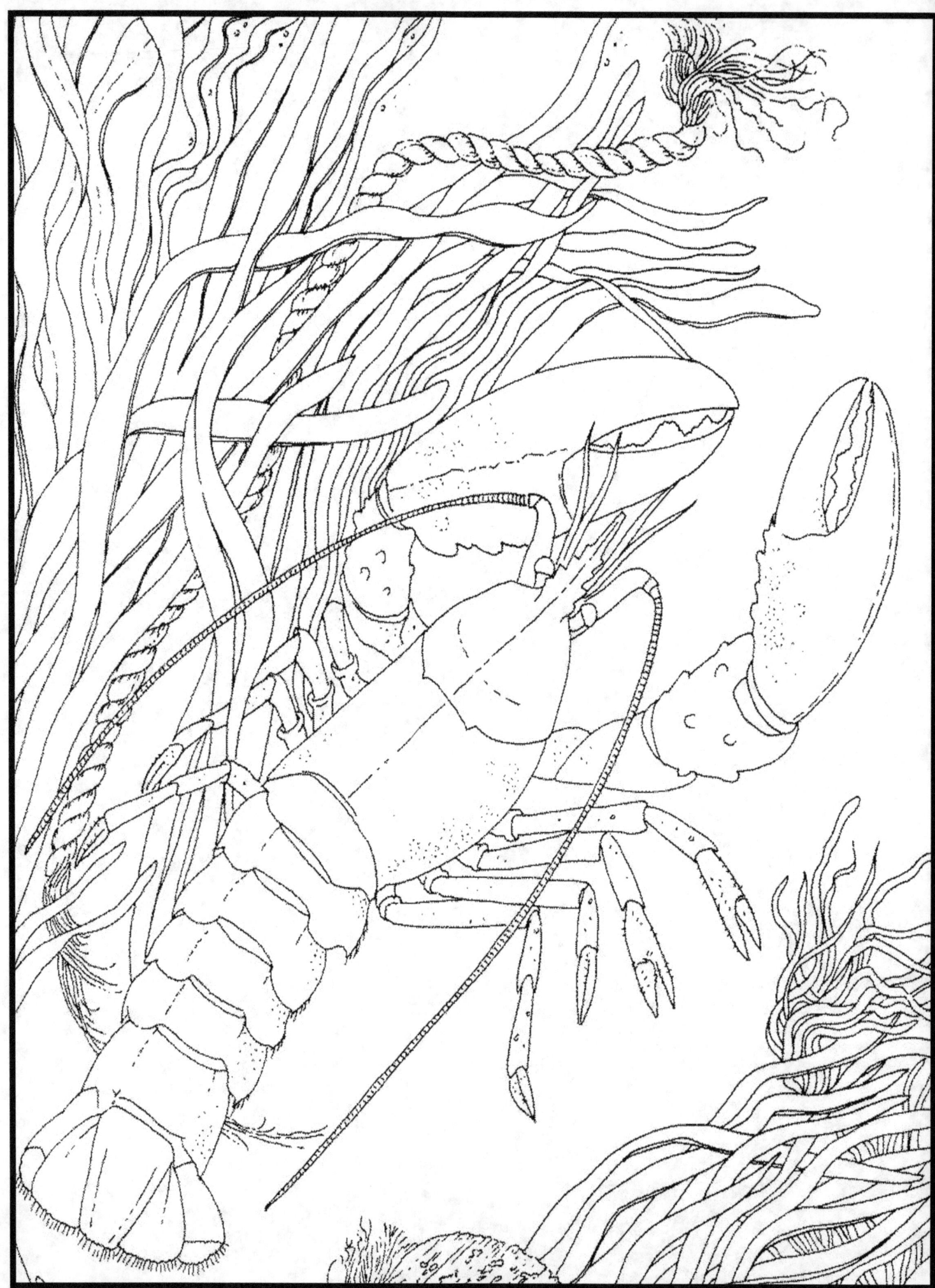

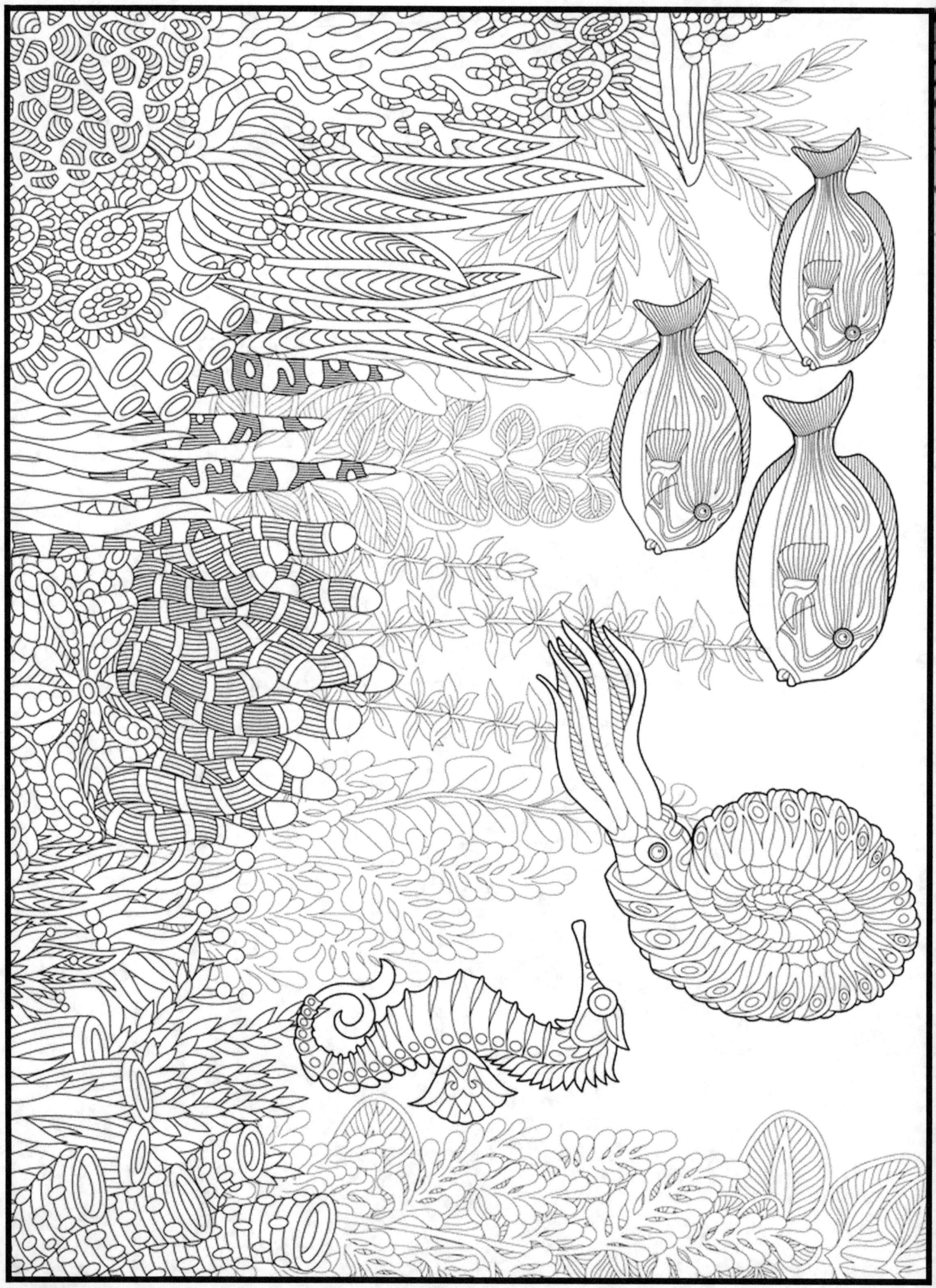

Found others coloring books

Relaxatio Amn

www.ingramcontent.com/pod-product-compliance
Lightning Source LLC
Chambersburg PA
CBHW060438220526

45465CB00008B/3189